LONDON'S CLASSIC BUSES

John A. Gray

Capital Transport

Front cover New RM 1657 and newly overhauled RT 3312 at Parliament Hill Fields in summer 1963. *Colin Brown*

Back cover Two CentreWest RMLs at Liverpool Street in 1995. *C.D.Jones*

Overleaf Panting for a grid start? Traffic lights at Two Waters, just by Hemel Hempstead garage, hold back RTs 3679 and 2157 eager to return their passengers homewards after working in Apsley's paper mills. In the early 1970s, RTs were still needed to run such works' services; in February 1976 RT 3679 departed for a Yorkshire scrapyard, with RT 2157 being deleted from stock in the following year. *Capital Transport*

First published 1999

ISBN 1 85414 210 0

Published by
Capital Transport Publishing
38 Long Elmes
Harrow Weald
Middlesex

Printed by
CS Graphics
Singapore

©
Capital Transport Publishing
1999

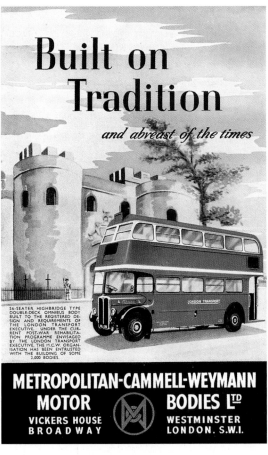

London's Classic Buses

'A pleasure indeed' was my thought when it was suggested I consider compiling an album portraying London's classic buses. It needed taking myself in hand, for the temptation to sit and recall times long past was frequently great.

There was no problem in defining *classic bus*. In the post-Routemaster era, the designs of the many double-deck bus types in London have been legion. This could indicate that no one of them is either designed to cope with the rigours of London service – or even to be in production, or service, long enough – to become a classic type. So 'classic' meant the front-engined, rear open platform bus worked by a crew of two, driving and conducting as a team. This concept was manifested in the AEC Regent Mk III and its Leyland Titan PD2 derivatives called here together the RT family, and the Routemaster.

Many photographers volunteered their slides and I thank them all. To reduce the offered number to those included was difficult, and certainly some would-be contributors will be disappointed by their exclusion. I am appreciative and grateful for the ready help in identifying the details – not always easy when our memories weaken – but any remaining errors are my fault. Particular thanks go to Graham Smith, David Mitchell, James Whiting and Sandra – still my wife despite my long interest in old buses.

To every reader, I hope the publisher and I, through these considered photographs, have done justice to the memory of London's classic buses.

Cambridge, February 1999   John A. Gray

The photographs accompanying the Introduction are by Capital Transport, C. Carter and Tim Clayton

Introduction

When London Transport's top officers considered and decided on a new standard bus in the 1930s, it was likely to be assumed that an open platform at the back would be preferred in a city bus design, and this arrangement seemed to dictate positioning the engine at the front. Earlier in the 1930s there had been two front entrance and two centre entrance double-deck buses designed for London work, based on the AEC Q type, but they never got further than trials. In this design, the engine was moved from its traditional front position to part way along the offside. In another effort, a modest batch of small single deckers was built with a rear engine; and in yet another design, some Green Line coaches were built on an underfloor-engined version of the Leyland Tiger chassis.

This new double-deck bus was to be a brand new design based on the experiences of operating a range of British-made buses largely of AEC manufacture but with considerable respect to Leyland and Daimler practice. What was seen as the best solution for London was put on the drawing board, transformed into RT 1 by AEC and LT's own Chiswick Works, and unveiled in summer 1939. Within a few weeks of its entry into service, the fear of war with Germany was borne out and, consequent upon the effort directed to equipping the military and the nation for war, fewer than half the original quantity of 338 RTs ordered were delivered.

Two years after the end of the war, deliveries of new RTs resumed, to a rather different specification, though the changes were mostly 'under the skin' and not seen by passengers. Some visible improvements were also made resulting from the experience gained in arduous operation for six years. It's a truism to proclaim that the RT set the standard for the quality-built urban bus. Despite that, very few were ordered by and supplied to operators in urban areas outside the metropolis. Reasons speculated for this include the points that the complete RT was relatively

expensive to buy; that it was needlessly sophisticated where passenger time and comfort were not so much of the essence, and bus crews were less demanding (or pampered, depending on standpoint) than London's. Further, it was engineered to be so thoroughly overhauled every four years or so that it emerged from works equal to a new bus. Provincial operators were not, and did not want to be, so geared. Added to that, they may have had a long wait for delivery, as the RT-type production was booked against London Transport's massive orders, which totalled nearly 7000. Thus London's first modern classic bus was born and introduced.

How is 'classic' defined? The RT type was in service for forty years, though not the same vehicles for all that time. Many postwar examples did however last for roundly thirty years. This was at least twice the expected age of their provincial counterparts. They lasted so long not because they were treated tenderly. They weren't. An individual bus could have five or six crews in 24 hours – and with all-night bus operation, a small minority of buses could be at work virtually continuously for one day/night/day – up to about 42 hours. An RT might cover, say, 150 miles in a working day, six and a half days a week (allowing for some Sunday and/or night work) for possibly 51 weeks a year, to give 50,000 miles a year, equivalent to one million miles in a twenty-year life. As London's buses at this time were not equipped with milometers, individual accurate records of distance covered could not be kept; but when proponents extol the merits of RTs by claiming many examples could have achieved upwards of one and a half million miles in their London life, they may not be wide of the mark.

As experience in the demanding arena of RT operation was gained, lessons were learnt to be recalled when the planning and design of the next generation was undertaken. The RT family had succeeded in ousting the remaining trams and all prewar, wartime and early postwar non-standard types by the mid-fifties, and before that thoughts had begun on the design of the replacement motor bus for the trolleybuses and, after this accomplishment, to succeed the RT family of buses.

And so the Routemaster. The Routemaster followed the general style of the RT but with differences – improvements were incorporated resulting from operating experience. One major difference did not result from RTs at all, but from some of the trolleybuses the Routemaster was to replace. This was the concept of having no separate chassis. Hundreds of London's chassisless trolleybuses had shaken over the East End's stone sett roads for a decade or more, much of which period had been when war damage produced even rougher road surfaces. The trolleybuses

were sturdy and withstood such punishment. The principle of strengthened bodywork carried on two subframes, one for each axle, was to be implemented. This meant saving weight, thus improving fuel consumption and producing economic operation.

Some improvements came from altered regulations, allowing larger 'box' dimensions. Although a start had been made in the late 1940s with both the postwar Q1 trolleybuses and the RTW type, both being eight feet wide, at that time special arrangements had to be agreed for each route to be worked by them. An easing of this situation meant that Routemasters could be of more comfortable width for operation everywhere. They could be

1ft 6ins longer too: enough to give an extra pair of seats on either side of the gangway on both decks, aggregating 64-seat buses as against the RT family's 56.

As the average laden weight of the RM was to be no greater than that of the RT (the absence of a full chassis and aluminium bodywork saving roughly the weight of the eight extra passengers), the same size of engine was adequate for Routemaster installation. Gear preselection was updated to simultaneous semi- or fully-automatic, giving a slightly greater acceleration; brakes were to be hydraulically operated rather than air, again saving weight.

Compared with the single prototype for the RTs, four prototype Routemasters (RMs 1 and 2, RML 3 and CRL 4) were put together. This level of effort suggested all should be well when production was under way. Disappointingly, it was not. The tale of the less than perfect RT brakes in early days is common knowledge, to cite just one RT lesson; the Routemaster was to provide rather more shortcomings. However, the will, the expertise, and the cash were available and utilised to put matters right and in due course, they were. The Routemaster was accepted as a most comfortable bus for passenger and conductor, and a considerable improvement, now with power-assisted steering, over the RT for the driver. The Routemaster was on its way to becoming a classic. To underline its achievement, the remaining vehicles have largely been refurbished and many re-engined (because new parts needed in the overhaul of the original units could no longer be obtained) to extend their lives even beyond the record of the predecessor RT. Some Routemasters still serving central London six days a week are forty years old.

Left After the prewar and wartime non-standard buses had all been withdrawn, by the mid-fifties the RT family ruled inner London without dissent. As last vestiges of wartime restriction and drabness were removed, people finally felt more assertive in their improved present and confident in their future. Visibly identified with the new optimism and stability in the London streetscene was the red RT, coming to be known by visitors – native and overseas – as the archetypal London bus. Both its ubiquity and convenience contributed to this impression. Unlike much later designs, its halfcab with bonnet side, subtly curving bodywork lines and upright back combined to produce a workmanlike but unaggressive demeanour, characteristics which were likely to have been given quite some thought when the successor, the Routemaster, was in design. Those best succeeding in crossing London Bridge on a fine sunny morning in May 1955 appear to be grandparents and grand-daughter: perhaps they've given up and alighted from the route 13 bus at the end of a queue of six RT-family buses heading towards London Bridge Station. RT 3965's conductor has time to talk to his driver, while in front of the then popular Morris Oxford saloon, RT 928's driver looks resigned to a continuing wait. All indicators are set properly except for those on route 133's RT 3965. *R. C. Riley*

Above At the next upstream road bridge, Southwark, in the same month RT 2495 allows a study of the former livery with light cream-coloured upper deck window frames. Registered in the KXW series, this bus was in one of the last batches to be so delivered from the bodybuilder in spring 1950 before the gradual change to 'all red' as depicted plentifully at left. This particular vehicle spent its first 15 months of life based at Chiswick for research purposes, delaying its first overhaul until shortly after this photo was taken; others of its batch were overhauled in early 1954. *B. A. Jenkins*

This fine study of Cravens-bodied RT 1498, standing opposite Southfields Underground station, exudes the June 1955 atmosphere, peacefully awaiting an exciting day at Wimbledon Tennis. Special services for annual events including Epsom Races continued to be needed. Norbiton garage's bus presents a clean and ready-for-work appearance, suitably informing potential clients of its duty. The bus stand, outside the period hairdresser's, is marked by the 'dolly' bus stop seen through the first window. *Bruce Jenkins*

RT 93 at Nazeingwood was one of eight 'prewar' RTs used in the Country area at the end of their service lives. Seven were repainted green and sent to Hertford garage for use on route 327, on which their weight, at three-quarters of a ton less than the postwar RT, allowed them to cross a weak bridge at Broxbourne. The bridge, spanning The New River (built in the 17th century as a fresh water supply for London's Finsbury district), was due to be strengthened, but until it eventually was at the end of August 1957, the seven soldiered on. The eighth remained red and spent only two short periods at Hertford, probably as a substitute for an unavailable green one. *Frank Mussett*

Displaying the distinctive 2RT2 appearance from the rear is RT 36. The rear roof box was used to show the route number for only a brief period at the start of the type's life before falling out of use, initially because of wartime restrictions. *Prince Marshall*

The policewoman strutting past No. 300 Vauxhall Bridge Road – Peabody House – pays no attention to Cricklewood garage's first prototype Routemaster, RM 1, as it heads southwards just beyond the former tram terminus for Victoria. Here the bus was on service trial in the summer of 1956. The attempted cream relief band around the front above the lower deck window line was abandoned in favour of the arrangement shown on RM 30 opposite, only to be re-introduced later and more successfully by halving the ventilator grille's depth. The attractive bullseye and dummy radiator arrangement at the front was short lived, owing to the need to improve engine cooling by moving the radiator from under the floor.
Geoff Morant

After trials with the Routemaster prototypes, production RMs entered revenue service in the summer of 1959. Allocated to Willesden, Riverside, Cricklewood, Hackney and Clay Hall in small numbers, these early operations were on a trial basis, all potentially working into London's core. On Golders Green Underground station forecourt the patterned stone setts support the weight of early production RM 30 and the slightly heavier Metro-Cammell-bodied Leyland RTL 770 both taking a rest from work on the same route. The Routemaster advertises nothing at the front: the posters were dummies, thereby avoiding blank panels. The all-brown wheels appear dull when compared with RM 1's, opposite.
Iain MacGregor

Left On its way from Liverpool Street to its home garage and passing Charing Cross station in evening sunshine, RTL 1607 carries a full load. Some passengers may stop off at Olympia's event, highlighted on the slip-board below the lower deck front facing window and just visible beyond the driver's duty card stuck in the lower windscreen. In this 1963 picture, the bus clearly suffered dents in its roof dome on both sides. *Fred Ivey*

Below left The RTWs came rather later to innermost streets; buses of such measurement were believed to be too wide for narrow or congested areas, until the 1950 trials showed the Metropolitan and City of London police authorities there were no insurmountable problems. The 74 was inner London's first eight feet wide motor bus route, Camden Town—Putney Heath, and for most of the life of the class central London became their operational territory. Reflecting at a wet Willesden garage in March 1958 stand RTWs 168, 429 and one other, demonstrating their ability in use from the outer north-west at Edgware on the Sunday version of route 18, to the East End at Old Ford on the Sundays-only 8B. *Ron Copson*

Below In November 1959, Poplar depot's trolleybuses were replaced by the first major introduction of Routemasters working the new bus routes. New RM 132, with single-pane upper deck front windows, passes under the since removed railway bridge over Ludgate Hill as it leaves the City on a quite short shuttle of the extended trolleybus replacement route 48 (Poplar-Waterloo), the section between St Paul's and Waterloo being given twice the frequency of the complete route. Until then it had been virtually unknown for a form of 'lazy' blinds to be displayed in the City. *Fred Ivey*

LONDON TRANSPORT
Bus-about tickets

Winter trees frame this picture of St Paul's Road, Brentford – the terminus once known as Half Acre, though that named thoroughfare is just around the corner – where MCW-bodied L3 class AEC trolleybus 1381 running an extra journey on route 667 overtakes RM 1009. Hanwell trolleybus depot, on conversion to Routemaster operation of the Uxbridge Road and Boston Manor Road routes in November 1960, assumed some duties on the then 97 with Routemasters supplanting Southall's RTs. The 667 succumbed to RMs as route 267 at the end of London's trolleybuses in May 1962. *Tony Belton*

Right In the days when the Hampton Court to Walton-on-Thames road was firmly in red bus territory, RT 849 flashes orange to pull away from a stop. One of London Transport's usual focuses on design detail was to provide a dedicated place for the driver's duty card in the cab door panelling. As many pictures show, it was often found convenient to lodge it in the gap between pipework and the lower windscreen frame. *Geoff Rixon*

Learners in April 1962 could have developed their driving abilities on either red RT 127 (above) or green RTL 1278, both photographed in Chiswick High Road. The 1940-built 'pre-war' RT 127, displaying a painted-on Upton Park garage code, certainly looks past its best days, now with top front number box panelled over. The bus was one of the few 2RT2s to have quarter-drop front upper-deck windows (thereby giving superior vision for front seat passengers!) and one-third drop elsewhere, later to standardise on RF single deckers and Routemasters. Red reflections from the Albion lorry are caught in the pristine green RTL 1278, here showing no garage code, but would have run out of Hatfield some two years earlier. It was there that the 18 green RTLs were used on trunk routes such as the 303 to New Barnet and 341 to St Albans for not much more than a year before returning to the Central area for use as learner buses. *Fred Ivey*

18

'Roofbox' RT 3507 returns to Sevenoaks with a light load when a Green Rover ticket gave a good day's travel value on the green buses for merely six shillings (30p). The bus left LT ownership as early as October 1963 as part of a purge of roofbox vehicles in the Country Bus area, where it was commonplace for staff to have to change the route number during a duty, sometimes more than once. Route 421 was one of the Country Department's shorter routes, to Kemsing and the hamlet of Heverham. Mention could have been made on the route blind of the traffic focus, Bat & Ball Station. *Chris Aston*

Left Some time before the new bus station in front of Hounslow's replacement bus garage was roofed over, RT 369 leaves Kingsley Road for the High Street and points south-west. Continuing to carry the low-valence roofbox style of bodywork as made for it, the bus sets out on a short 117 journey that will traverse only the last two places mentioned on the route blind, and then in reverse order. *Fred Ivey*

THIS BUS is one of London Transport's latest Routemasters. Its aluminium body has been left unpainted so that it can be compared, in terms of wear and maintenance, with those painted in the standard red livery.

Londoners and visitors are invited to write to the Public Relations Officer, 55 Broadway, S.W.1, saying what they think of it.

To see how well unpainted aluminium would wear, RM 664 was left that way, apart from painted wheels, until 1965. Its transfers for fleetname and stock number could have been improved had the red version been used as on private hire single deckers some ten years earlier. The bus here was working from its intended garage – Highgate – but not on its intended route, instead being caught at King's Cross on the 17. The basic simplicity of the internal poster, using London Transport's own 'Johnston' typeface as to be expected, perfectly complements the bus's design. *Fred Ivey*

LONDON TRANSPORT
55 Broadway, S.W.1.

4.7.61   TPW.1652

"SILVER" BUS TO BE TRIED OUT BY LONDON TRANSPORT
EXPERIMENT WITH UNPAINTED ALUMINIUM FINISH

An "all-silver" bus with an unpainted aluminium alloy body is to be tried out on the London streets, the London Transport Executive have decided. It will be allocated to Highgate garage and, starting shortly, will run through the West End on route 276 (Brixton – Whitehall – Regent Street – Camden Town – Tottenham).

The bus, RM 664, is one of London Transport's standard Routemasters which already have lightweight aluminium alloy bodywork, but normally are painted in London Transport's standard red livery.

Except for the wheels, and the gold London Transport insignia on the sides, the experimental bus will be entirely silver in appearance, certain moulded plastics parts being painted silver to match the unpainted panelling. It will run in ordinary service so that its actual performance on the road can be compared with buses in the normal painted finish as regards wear, maintenance costs, and appearance.

"We want to try out an unpainted aluminium bus to see how it stands up to hard daily service, how it looks in traffic, and how the public like it" a London Transport official said. "We expect that it will cause some controversy and we shall invite the public, by a notice on the bus, to write and tell us what they think of it. This is an experiment, one of many we make, and it does not mean that we intend to change over to silver buses generally.

"Some of the staff have already christened the unpainted aluminium bus 'The White Lady'".

On Saturdays, the bus will run on the Victoria Station – Edmonton section of Route 127.

PRESS OFFICE
ABBey 5600: Ext.70

Gordon's Stands Supreme

NELSON
for better taste
The tipped cigarette
with the fine Virginia flavour
3'10 FOR TWENTY

17 HIGHGATE KINGS + GRAYS INN ROAD ELEPHANT
NORTH FINCHLEY

17

RM664

LONDON TRANSPORT

WLT 664

The RMC class was the first production run of a double decker designed for Green Line work, ironically coming at a time when Green Line had begun a decline in usage. Based very much on prototype CRL 4, RMCs 1455 and 1457 were photographed on the finishing lines at Park Royal Vehicles, while 1487 leaves the works with the Grand Union Canal bridge in the background. The new vehicles began to enter service from August 1962 and routes 715 and 715A were the first to receive them. RMC 1462 is pictured at Hammersmith a few days after starting work. *Capital Transport, Fred Ivey, Colin Brown*

Route 230A was an odd by-product of the plan to introduce RMs to replace RTs on a '10 for 11' basis to take into account the RM's 15 per cent higher seating capacity. In October 1962, Harrow Weald was to have received RMs for most of its RT routes and the 230A would have made use of some of the staff made spare. In the event the busmen refused to accept the RMs on a reduced-frequency basis. However, the 230A was introduced anyway, serving a number of new roads including Kenton Lane where RT 1946 was caught at Belmont on the route's first day overtaking a splendid AEC lorry involved in renewing the road surface. *Fred Ivey*

One of the RMs whose entry into service was slightly delayed while London Transport was forced to revise its plans for RT replacement is pictured at AEC's Southall Works in November 1962.

As things turned out, RMs started to replace RTs first on routes penetrating central London, except for one, the 37, an inner southern ring route. In December 1962 and the early part of 1963, large batches of new RMs enabled the withdrawal of similar numbers of RTs, mostly roofbox ones since these were generally the oldest and due for withdrawal. Park Lane carried the greatest flow of Routemasters at the beginning, on routes 16, 36 group and 73, which last route in Knightsbridge was joined by Mortlake garage's on the 9 in April 1963; the western part of Oxford Street enjoyed RMs on the 13 as well as the 73. A classic bus in the making had begun to usurp its established predecessor, a process which would never be completed. In near original condition (BESI scanning panel forward of the Pickfords advertisement was an additional detail), RM 1551 picks up in Hammersmith when route 73 had no thought of going to Victoria. The streetscape here at Hammersmith Bridge Road has been changed totally in the last thirty years with the demolition of the ABC cinema and widening to dual carriageway. *Colin Brown*

The 36 group of routes began to receive RMs in the harsh January of 1963. Rye Lane garage's RM 1420 was photographed during the changeover period at Queens Park. The front posters are another example of London Transport Advertising's obscure messages to fill unsold poster spaces. *Fred Ivey*

Below Built for trial use alongside RMLs on route 104, RMF 1254 never entered service on an ordinary bus route in London. After a demonstration tour of operators outside the capital, the bus was used on the service between West London Air Terminal near Gloucester Road and Heathrow Airport.
Iain MacGregor

Clapton garage's hard standing had ample space for the three RTLs shown here: left to right – 494 with Park Royal bodywork, and 727 and 892 with the apparently sturdier Metropolitan-Cammell Carriage & Wagon Co's product. The immediate distinguishing feature between the two makes was the depth of the tween-deck cream band – shallower and with more prominent beading above on the Metro-Cammell version. *Fred Ivey*

A few route number roof box RT10 bodies were mounted on overhauled Leyland chassis when it had been decided their London life was to be cut short. Walworth garage's RTL 73 rests at Vauxhall Bridge Road terminus, Victoria, in October 1965 on route 185 which had replaced the circuitous tram route 58 fourteen years earlier. Despite proposed early withdrawal, another four and a half years' LT ownership lay ahead for RTL 73. No inspector seems to be around to answer the wall mounted LT telephone . . . *Capital Transport*

The Daily Express chose an apt slogan for its illuminated advert on RM 1987, seen here at Hyde Park Corner. London buses seemed less attractive to commercial advertisers in the 1960s than they were to become in the 1990s, and illuminated panels at the front of the lower saloon and on the offside, as shown here, were seen as a way of drumming up more business. Their night-time illumination was very effective and for a few years the panels proved popular. While several passengers observe the photographer, another begins descending the stairs. The driver looks in the nearside mirror, in anticipation of two rings on the bell. *Colin Brown*

Wide ones compared in May 1965 at Hackney garage. Tidying the front of Routemasters continued in the mid-1960s as RM 2147 shows, and compared with RTWs 352 and 497 is considerably neater, though changes were not to stop there. The standard of driver's forward vision is much the same on the two types, for although the Routemaster's engine is set lower, the windscreen stays at about the RTW's depth and in the same position; the horizontal subframe is lower on the RM. Openable windscreens can be a mixed blessing; when opened in hot weather the cool draught admits insects, yet when closed afterwards, their poor seals can leak in the next heavy downpour. *Capital Transport*

Top In the mid-1960s, the Elephant & Castle district lost forever its previously popular shopping and closely-packed residential locality. Always a centre of traffic congestion, the new roundabout was here under construction in the belief that movement would be easier. RTW 285 appears to be proving the point. *Capital Transport*

Above The established gyratory traffic scheme at Marble Arch is the scene of this view of RTW 215, carrying both an early and a later design of BESI (Bus Electronic Scanning Indicator) plate for bus tracking purposes. The 74B was one of six routes so equipped. *Geoff Rixon*

In May 1965 the inspector's desk on the steps rising into Broad Street station is a practical period piece, with adjacent wall mounted telephone to hand, its little door ever open. All gone now: Broad Street station's quiet caverns have been replaced by an impersonal subterranean shopping mall and endless offices rising disturbingly above; inspectors by cab radios, and RTW 13 by a Routemaster. The 11 now terminates in a new, constricted, turning area one hundred yards eastwards illustrated on the back cover. *Capital Transport*

The two Green Line versions of the Routemaster compared. Normal length RMC 1506 (*left*) was last of a batch of eight coaches allocated to Watford's Garston garage in November 1962 for use on route 719. It was photographed in Watford's one-way system in its original condition, before cosmetic modifications were made to bring the sub-class more into line with the later-developed version, as RCL 2252 (*below*). While the 68 RMCs seated 57 passengers, the 43 longer RCLs had seats for eight more. The larger 11.3 litre engine powered the longer and heavier coaches, so as to allow similar performance characteristics to the shorter RMCs with the standard 9.6 litre unit. In keeping with all production Routemasters, bodywork was built by Park Royal Vehicles. *Capital Transport, Colin Brown*

Well before August 1969, when this photograph was taken in South Street, Romford, Green Line RTs had been supplanted by coach Routemasters on their designated routes and so Grays garage's smart Regent now runs the trunk Country route 370 though still in Green Line livery. Indeed the Green Line RTs were the standard bus design: no deeper seats or parcels racks as had been conceded on pre-war single deckers. Devoid of all advertising and with a pale green band together gave that touch of distinction still evident here. Meanwhile red sister RT 3747, behind, will venture no further than the more prosaic Upminster Park Estate on the 193, a route number nowadays bearing little similarity to this predecessor. *Capital Transport*

An old upper case route blind graces newly flake-grey banded RT 495, yet new lower case is shown by old cream banded RT 2141: such are the ironies when standards are changed. Both Muswell Hill garaged buses look in their prime as they take a break amid late nineteenth-century Dutch-style architecture at the Hill's Broadway, May 1967. A third RT's route number roof box can be made out through the 134 bus's upper deck windows. *Capital Transport*

Brand new RML 2301 pulls in front of a similar Central sister in October 1965 at Godstone garage. The red buses were on loan to the Country Department as insufficient green RMLs had been delivered. Posters were affixed to the window next to the entrance assuring passengers that Green Rovers were available on the bus. *Capital Transport*

Right Giving green relief to an otherwise seemingly all-red full-sized West Croydon bus station, brand new RML 2315 pulls out bound for its home at Godstone with a sizeable load. Seeming to set the standard for Country bus presentation, it's hard to find a fault: advertising posters are properly placed, destination display accurately shown, and even the running number stencils are slotted squarely. *Capital Transport*

On the old Hertford–Hatfield road, RT 3442's driver allows the cyclist plenty of room. Trunk route 341 served a terminating town's suburb, in this case the destined Marshalswick area of the city of St Albans, as traditionally did several Country department routes. What better way to travel through such scenery? *Dave Brown*

Farningham's by-passed narrow main street unusually accommodates a bus stop which appears to be erected in a front garden. RT 3162 exchanges passengers at the stop shared by all three road services in July 1969: red route 21A from Eltham, premier green route 401 and Green Line 719. *Dave Brown*

COUNTRY BUSES

MAP AND
INDEX TO
PLACES SERVED

*Specially folded
for easy reference*

LONDON TRANSPORT

55 BROADWAY, S.W.I

01-222 1234

1968/69

A deserted request stop at Sevenoaks Weald jumps out from 1969 summer's foliage to be ignored by RT 4505 intent southbound on the twisty 454, a route that ran entirely within Kent. The bus lasted with its later LCBS owner longer than many, not going to Yorkshire for scrap until spring 1975. *Dave Brown*

Left During the 1990s, an eventually unsuccessful campaign was mounted to save the closed St Albans garage for conversion to a transport museum. One exhibit may well have been an old inmate similar to RT 650 here at Marshalswick on the city's local circular route 354 during the 1962–63 winter. The road surface looks to be positively slippery – highway authority application of sand/salt mix was neither so universal nor effective as it later became. Judging by the snow and icicles, the bus had probably been left outdoors all night; this garage's allocation could not all be parked under cover. *Colin Brown*

Above Youths ignore the red appendages on Dorking's compulsory stop M qualifying it for use only by journeys terminating at the garage quite some distance ahead; RT 4048 will not collect them on its way there after a long journey from the Chelsham area and through the background North Downs in April 1969. *Dave Brown*

Old Stevenage was very wet and deserted after a January 1966 cloudburst. RT 3912 draws away from the High Street stop alongside the Great North Road, A1, when it still came through the town. New routes in the post-war towns had used up all 3XX and 4XX numbers and so 8XX and 85X numbers were introduced to continue the Country area's north and south sequences respectively. *Peter Thatcher*

St Mary's Square in Hitchin was later to succumb to private transport's storage needs as a car park, but before that in July 1966 it contained little other than a patient RT 3422, gradually accepting passengers whose ease of boarding is evident. Route 303 was an LT classic itself in a way, going through no fewer than three of the six new towns served by London's buses, and terminating at a useful interchange with trains for King's Cross and the Central bus network. Red finials atop stop posts graced both Central and Country bus stops; green ones completed Green Line coach stop posts; and one above the other adorned posts where both buses and coaches stopped. *Peter Thatcher*

Under storm sky and somewhat remote from the advertised Victoria Line stretch now opened from Walthamstow, another loaned RT, 2090, takes a break at a less frenetic Bookham Station in British Railways Southern Region days (though the station's lamp tablet is an earlier Southern Railway design). The bus will regain familiar frenzy on returning to Kingston. Leatherhead's own green RTs were drying out in September 1968 after the garage had been inundated by River Mole flood water. *Capital Transport*

Though delivered new in 1953 as a red bus, by 1966 RT 4563 had become green, here at Uxbridge Station terminus in Bakers Road, working out of Garston garage on the trunk 321 route, whose section southwards beyond Maple Cross eventually withered. Red sisters work the local 204A via Hillingdon Hospital (RT 2484) and the 223 to Hounslow. *Capital Transport*

Tring entertains home-garaged RT 3238, originally a Green Line vehicle from Romford (RE) garage, still looking sprightly in July 1969 while entrusted with the northern Country area's premier trunk route. The 301, in common with the 341 illustrated earlier, was another to delve into terminal suburbs, this time a choice between Watford Heath or Little Bushey. *Dave Brown*

Left and Right 'I've been driven all across North America', proclaims RT 2776 as evidenced by the 'bookends' – extra ventilators in the front dome – and the plaque inside. A rectangular scoop beneath the canopy for extra lower deck ventilation had long since been removed by the time of this June 1965 photograph at Wanstead and it is surprising that the upper deck ventilators survived until the end of the body's life on RT 1708. Route 101 had its arduous times, but distances aren't quite so great as those in that fondly remembered 1952 adventure in company with RT 2775 and RTL 1307. A weak mechanism and a strong wind occasionally combined to produce a useless windscreen wiper, needing an athletic input by a driver to re-instate, as would have been necessary here. *Capital Transport*

Below With its nether area looking slightly reminiscent of the earliest London trolleybuses – the 'Diddlers' – RM 738 shows its experimental engine frontage in June 1969 evening sunshine while driving round the block in terminating at Woolwich Arsenal station. One of two dozen RMs fitted experimentally with noise-absorption material in the engine compartment, this bus was alone in going further in pursuit of quietness with these ugly baffles, which made little difference to sound levels but could have prejudiced engine cooling. Route 99 is rare in remaining much the same over the last 40 years. *Bill Cottrell*

With red-painted grille, RM 1039 observes the compulsory stop (including – in black – Red Arrow route 500) outside the recently completed Hilton Hotel in Park Lane in September 1966, when an all-day everywhere bus ticket – Red Rover – cost six shillings (£0.30). *Capital Transport*

Right FRM 1, modified from original condition by the fitting of opening windows following its fire, turns into Tottenham garage at the northern end of route 76. Outside of London by the mid-sixties the rear-engined bus was replacing the traditional front-engined double decker as a matter of course. *Capital Transport*

A NEW GREEN LINE SERVICE
727 LUTON STN - WATFORD JUNCTION STN
HEATHROW▸ - GATWICK - CRAWLEY
◆ FAST NORTH-SOUTH ROUND-LONDON ROUTE
◆ NEW ROAD-RAIL-AIR COACH LINK
Ask any Green Line Conductor for a leaflet

Kew   Richmond
Ham   Kingston
Surbiton   Hook Road
Copt Gilders Estate
65A
CHESSINGTON ZOO

LONDON TRANSPORT

RT 3896

LLU 695

97

RUISLIP ROAD
GREENFORD
PITSHANGER LANE
EALING · NORTHFIELDS
BRENTFORD

OPEN DAILY

BATTERSEA FUNFAIR
OPEN DAILY

TRUBROWN brewed by Trumans

LONDON TRANSPORT

RT 1699

KYY 526

Facing page Sunday route 194C entertains fresh RT 2787 on a
quiet spring day in 1968 at Croydon Airport terminus, some
five years before the end of the bus's London life.
Capital Transport

Above Classically posed: Kingston garage's RT 3896, fresh
from overhaul, heads southward through Surbiton on the 65A
in June 1967. Routeing stability south of Hook has never
remained for long; introductions, variations and cessations of
the 65A, 265 and 71 have all occurred down the years, with
main route 65 lasting the longest, only to give in to an
extended 71, though south of Chessington's reworked zoo,
even that has succumbed to a Country son, route 465.
Geoff Rixon

Left A third RT with a fresh coat of paint, this time Saunders
bodied RT 1699 at Northfields just after overhaul in August
1965. The overhaul was its last – roofboxes in London service
had just over five years left. A few RTs were allocated to the
otherwise Routemaster Hanwell garage, formerly trolleybus
depot; beforehand route 97 had run out from Southall garage.
Colin Brown

Right The year 2001 seemed a long way off in May 1968 when RML 2686 was photographed at Hayes End with adverts for the late Stanley Kubrick's film. At the time it would not have seemed possible that RMLs would still be in normal service on some of London's busiest routes on the other side of the new millennium. Earlier in 1968, London Transport had taken delivery of its last Routemasters, having placed its faith in one-man operated single-deckers which began to enter service in large numbers in September. *Capital Transport*

Below Uxbridge garage had a small Sunday-only RT allocation on the 207, on which RT 1929 is seen at Ealing long before registration numbers like the one it received on delivery in 1950 were worth big money. *Colin Brown*

RTL 1602 looks good for many years ahead and rather smarter than the surroundings of Golders Green Underground station in September 1968. By this time, the remaining Leyland RTs were employed only on routes 176 and 226 out of Willesden garage, and here the bus is destined for a brief but twisty journey on a 226 short working. No fleet number is shown on the cabside: the front corner panel is a replacement in undercoat. The end of the type in London service was by then well in sight; it came at the end of November. *Capital Transport*

Route 253 took over the busy and frequent arch-shaped Aldgate–Tottenham Court Road trolleybus 653 in February 1961 worked by Highgate garage, formerly depot. Here, Stamford Hill garage's RM 837 exchanges passengers at Stamford Hill some eight years later. Though much Routemaster detail had been modified by now, at least two items remained as when new in this case: the engine grille style, and the larger size all-capitals route blind. *Capital Transport*

Left Pioneering a new form of advertising, RM 1737 heads past Westminster Abbey with the first all-over bus advert in the country in 1969. More followed in London and elsewhere, and other companies to use Routemasters in this way included Yellow Pages, Ladbroke's, Esso, Barker & Dobson, Dinky Toys, and the now-defunct London Evening News. Brook Green Hotel, just on the Shepherds Bush side of Hammersmith Broadway, was at that time the western extremity of popular route 11, since attenuated at Fulham Broadway. *Geoff Morant*

Above Being subsumed into the huge Dutch Phillips electrical concern came later for this British electronics firm, but in 1973, Cambridge-based Pye was showing confidence in adorning RM 682, caught here in the outer north-western suburbs at Edgware. *Capital Transport*

Delight in them or loathe them, overall advertising liveries have 'enjoyed' sporadic periods of popularity. Their peak was reached in the early 1970s and six examples are illustrated at work at Trafalgar Square (Vernons), Becontree Heath (Myson), Barnes (Unipart), Willesden Lane (Ladbroke's), the Commercial Road (Evening News) and Holborn (Rand). The flashlight view of RML 2302 amply illustrates the final neatness of the bottom frontal area of the Routemaster, notwithstanding which drivers could claim difficulty in sighting the nearside extremity when compared with the RT. *Steve Fennell, Capital Transport, Colin Brown*

Left Looking every inch workmanlike (many of London Country's RTs were overhauled at LT's Aldenham Works in 1971–72) but sadly doing little business, RT 3249 negotiates Watford's constricted town centre on the long-established 302. RTs had operated the route continuously since their initial allocation to Hemel Hempstead (Two Waters) garage in the 1948 summer, replacing exhausted STs. When new in 1950, RT 3249 was one in the batch of 36 RTs for Green Line work out of Romford garage. *Capital Transport*

Above RT 3752 was unique in being repainted into the revised, slightly lighter, green shade adopted by London Country in 1972. It was also the last RT to be overhauled at Aldenham and lasted in service until 1977. *Capital Transport*

Left The open roundel on RM 420's staircase side (at Blackfriars) was applied to only one hundred Routemasters on overhaul. The device had been introduced for use on the rather unloved DMS type Daimler Fleetline double decker. *R.C. Riley*

Above An offside view of the unique rear-engined Routemaster, standing on the Potters Bar garage forecourt in autumn 1973. FRM 1 had been overhauled during the spring and summer, when, central in the front panelling, it gained an air vent where the previous open roundel had been. A filled roundel ('bullseye') had been introduced on to the sides of London's red buses, and FRM 1 had them too. Here it is ready for the 284 road, an undemanding local service where, no doubt, the single-doored and single-operator bus would cope more than adequately. *Steve Fennell*

RT 3679 was one of many of its class to continue working from Garston garage beyond its expected time for withdrawal. Routemaster family vehicles should have been on duty but were largely unavailable through lack of spare replacement parts. As if to underline the recall, Routemaster blinds provide the route information. *Capital Transport*

It's race day at Epsom, and anything goes, including one of the four 1977 RT recertifications (three of which were reliveried into National Bus green and white). Chelsham's RT 3461, weighed down with happy platform riders, gives support to a Euro concept of all-enclosed buses to those with a nanny mentality. *Graham Smith*

Right 'Green Line coaches to London' reads the scarcely eye-catching poster on the front of RML 2353 as it heads out of Butterwick bus station, Hammersmith, for Windsor. London Country's Routemaster coaches had all come off regular Green Line work by 1972; the long Routemaster buses were called in to run reliefs on routes such as 704 and 718 out of Windsor garage at busy times. Passengers look pleased at the prospect of a pleasant ride in the time-worn bus – only the stump remains of the nearside wing's marker post. *Colin Brown*

Angel Road, Edmonton, before widening as the North Circular Road later demanded, entertains RT 3321 on the long-established northern arc route 34 from Walthamstow, and another behind on the weekdays only 191 from Chingford. The 34's driver has made a half-hearted attempt to use the lay-by for the bus stop from which he is pulling away. *Capital Transport*

RT 689 is in the Uxbridge Road, Acton Vale, on what must be quite a hot day judging by the two men stripping at the bus stop opposite. A number of RTs had their front windows locked closed, including no doubt this one as all the side windows are fully opened. A Routemaster could have been the expected route 266 bus, but the 56-seater has more than enough room for the passengers carried on this occasion. Starch Green, last on the route blind, is an otherwise lost name for the Askew Road area. *Capital Transport*

Becontree Heath in 1976 still had a classic array of street furniture including lamps and a siren of the air raid warning type, to RT 4163's offside. The Weymann-bodied bus had been green and was eventually withdrawn for scrapping in April 1978. The well-known U-shaped route 87 went on to be broken into three separate services. *Capital Transport*

Above A pre-war street lamp still stands high above Park Royal-bodied Swift SMS 416 at the Station Parade stop in South Street, Romford, as RM 1426 comes by and finishes an infrequent timing over the 175A and RT 2717 departs stop X opposite. Both RT and Swift have flake-grey relief, whereas the overhauled Routemaster has white relief and roundel. *Capital Transport*

Far left RT 848 hastens past a DMS Fleetline along Yeading's dual carriageway in August 1975. Recently repainted, the bus gives the approximation of how it looked when newly into service nearly 27 years earlier at Elmers End garage. The original cream relief band is now pure white after up to a ten-year spell of flake grey; rear wheel discs have gone and trafficator 'ears' are additional. When new, the bus numbered RT 848 would have had its upper deck window frames finished in matching cream, and carried reduced route information. This sprightly pensioner made it into the next decade in LT ownership, though by then was no longer active. *Capital Transport*

Left An early sunrise in June 1975 greets the Commercial Road and RT 779 homeward bound on a night route. Barking garage's breakfast beckons . . . At this time, the area north of Shadwell (behind the contemplative conductor) still showed evidence of temporary repairs to buildings after World War II bomb damage. The bus is properly dressed for its empty journey: even the route number is shown correctly beneath the canopy. RT 779 went to Yorkshire for scrapping some 15 months later. *Capital Transport*

All told, 27 Routemasters were repainted silver and red in connexion with H.M. Queen Elizabeth II's silver jubilee in 1977, 25 of them for actual service and two others to drum up sponsors. They looked very smart down to the fleetname substitute slogan 'The Queen's Silver Jubilee London Celebrations 1977'. Uniquely the buses were numbered temporarily into a new series, SRM 1–25, for their silver period. Here, SRMs 21 (RM 1870), 15 (RM 1903) and 23 (RM 1902) stand at the eastern end of Hyde Park's South Carriage Drive, itself unusual since traditionally, any commercial vehicles displaying advertisements were banned from the royal parks. *Capital Transport*

Amey Roadstone Corporation sponsored RM 1889's silver livery to become SRM 4 for the Queen's Silver Jubilee London Celebrations, though the company's repeated initials at the side poster position lose impact through being scarcely readable. The bus pauses at Gray's Inn Road junction with Holborn in the days when route 25 negotiated the expensive Bond Street area on its way from Victoria. *Capital Transport*

While SRM 22 (otherwise RM 1900) passes Hyde Park Corner nearest to the Duke of Wellington's imposing Apsley House, equally smart early-bodied RM 705 comes up from Victoria via Grosvenor Gardens on the offside. Doubtless recently overhauled, the bus wears arguably the most favourably remembered of red London livery versions down the unending Routemaster years. *Fred Ivey*

Right Towards the other end of the scale, poor old RT 1161 looks most of its advanced years while substituting for a Routemaster in Piccadilly. During 1976 ten garages were each allocated a few RTs to alleviate Routemasters' lack of availability caused mainly by a severe shortage of replacement parts. Clapton's RT uses a Routemaster's side route blind in its large front box. *John Miller*

Left Shoplinker RM 2207 heads southwards along Kensington Church Street chased by another RM on route 52. This road was involved in the 1950 trials to see whether RTWs with their 8ft width could safely negotiate *inter alia* Notting Hill Gate, thus opening the way some years later for universal Routemaster availability. *Capital Transport*

Below and Right The veracity of the poster statement on RMs 2142 and 2130 could be called into question since the advertiser did not exist in 1812 – but by 1829, George Shillibeer's Omnibuses had been introduced into west London and they were believed to have worn a livery in the style of these RMs. Twenty five Routemasters had a sponsored repaint to commemorate the sesquicentenary of Shillibeer's Omnibus. *Colin Brown, Mike Harris*

All dressed up with somewhere to go, RT 422 makes a routine trip on the 94 on the route's last day of RT operation in August 1978 – at the end of the day it performed the last journey. The 94 had been continuously worked by RTs for 28 years. Such decoration was not uncommon when RT and RM routes were being converted to more modern vehicles, though in this case the vehicles taking over the following day were RMs.
Steve Fennell

The final run of the RT on Saturday 7th April 1979, photographed through the upper deck emergency exit window of the RT ahead and showing Weymann-bodied ex-Green Line RT 3251, which had been bought back by LT from London Country Bus Services six and a half years earlier. RT 3251 was meant to be the last official bus in this post-service cavalcade over route 62, but beyond the next (official) bus behind follows a procession of preserved RTs – one with cream upper deck window frames, a green one and three more red ones, the last having just left Barking (Fair Cross) garage to face the cameras here in Longbridge Road. *Capital Transport*

By August 1980, when this photograph was taken, RCL 2240's appearance had been quite transformed along with 25 others of the type in a thorough Aldenham overhaul which modified them to a model of clarity and comfort for city bus work after their long decline from Green Line coaching. Back then, the lady passenger wouldn't have had an open doorway through which to look for her stop. *Geoff Rixon*

On the Royal Wedding Special Tour, heavily-laden RM 520 follows the royal couple's processional route from the profusely decorated Fleet Street across Ludgate Circus and then uphill towards St Paul's Cathedral one August 1981 afternoon. Such liveried buses were to be seen as red parcels tied by silver ribbon whose bow is below the lower deck windows. *Geoff Rixon*

Believed to be the first occasion a long Routemaster ran on route 65, smart RML 2711 accelerates away from a mini-roundabout, the bane of a bus driver's life, on 5th July 1982 in Surbiton. *Geoff Rixon*

Below left RM 254 had become a 'showbus' by the time of this February 1985 picture. Items now reverted to original style include the framed registration number, the wing grilles beneath the headlamps and the full depth ventilator intake intruding into the cream band. Was any passenger misled by route 71's blind for Richmond Hill, Sandy Lane and Tudor Drive? *Geoff Rixon*

Right Largely restored to original state, RM 1563 leaves the erstwhile Butterwick bus station at Hammersmith in April 1981 to regain entry to a traffic jam intent on Fulham Palace Road and Hammersmith Bridge. This none-too-easy manoeuvre, even at lower traffic flow, is now thankfully obviated by the replacement bus station built into the Underground station; though even here there's a set of traffic lights to control bus movements within the approach. Until Mortlake garage was closed, its staff had a deserved reputation for turning out their buses very presentably, an effort carried to the ultimate here to include two kinds of Routemaster 'radiator' badge on what was for a time a 'showbus'. *Geoff Rixon*

Riverside Garage's RM 1091 passes a side entrance to Victoria Station in Buckingham Palace Road. The body on this bus has non-openable upper deck front windows of the kind that equipped the earliest Routemasters. The bus celebrated London Transport's golden jubilee year, 1983, freshly repainted to include a simulated gold between-decks relief band, inevitably lost on passers-by. Its smart appearance was a little spoiled by the application of paper notices to the inside of the upper deck rear side window – a growing trend more recently. *Geoff Rixon*

Right Golden bus RM 1983 celebrated the LT jubilee in 1983 uniquely in this livery that contrasts with the street scene on the inner suburban route 253. The bus was in turn operated out of a dozen or so garages in the spring and summer though mechanically it was regrettably unreliable and so missed one or two appointments. It ended up at Clapham in September till the year's end. *Steve Fennell*

Detail changes have been manifest down the Routemaster years, quite as many
if not more than the post-war RT types endured in their operational 32 years.
The four posed buses adorned under garage initiatives with their interpretations
of the 1933 style livery show further detail differences. In front, RM 1933 was
Chalk Farm's clear choice; Sidcup's RM 8, Seven Kings' RM 2116 – fortunately
since privately preserved – and RM 17 from Willesden. The roof protrusion
housing radio equipment shows well – all four have it in common. *Geoff Rixon*

RM 1278 during a very short-lived all-red livery trial collects customers on a warm October 1988 evening from outside the former St George's Hospital at Hyde Park Corner, now the Lanesborough Hotel. The bus shows not only the red/ orange/white roundel adopted after the white bullseye motif, but also a late and not very readable attempt at upper case lettering for route points. *Geoff Rixon*

Pictured in Ealing, decapitated Routemaster coach 1510 loses dignity with scratch blinds, a plastic grille ornament marking a Comic Relief Red Nose Event and an inappropriate poster while working a special journey on Friday 4th June 1993 in connexion with an Alperton garage open event held the following day. Its Westbourne Park garage code is covered by the running number plate from Alperton. *Gerald Mead*

Left Kentish Bus, previously London Country South East, introduced its attractive maroon and ivory scheme on its Routemasters working the rather shortened route 19, clearly detailed on each side of the buses, whose appearance was heightened by the absence of commercial advertising. RML 2452's driver has taken full advantage of the openable windscreen while driving along Shaftesbury Avenue. On a point of subtle detail, the grille badge restates the dedication to route 19. *Capital Transport*

Right In this view route 13 from Aldwych is in the care of BTS (formerly Borehamwood Travel Services), whose leased RML 2487 heads northwards along the Finchley Road bus lane. BTS's livery is a touch on the orange side of LT red, supplemented by the deep yellow relief band. Detailed changes from original include blanking off the between-decks front ventilator intake, substitution of fixed trafficators for flexible 'ears', reflective registration plate, and company logo heading the grille triangle.
Capital Transport

By now minus roundels, sister long Routemasters 890 and 883 come round from Oxford Street to circuit Marble Arch while operating from Upton Park and Camberwell garages respectively. Both have similar frontal modifications, neither now foglamp-fitted, though 883's hemline still shows clearance for one. The fixed front trafficators at cab ceiling level are found to be more satisfactory than the previous rubber-framed 'ears'. A little further back on the relief band of both vehicles is the loudspeaker grille for the alarm system, another modern addition. *Stephen Madden*

Right Upper RMs on route 139 retained original decor and seat moquette pattern up to the time the route was converted to one-person operated Darts in 1998. Some RMs still run in this condition, forty years after the production vehicles entered service. This photograph was taken in Baker Street shortly before the conversion of the 139. *Capital Transport*

Right A detail view in Oxford Street of RML 2307 in December 1997 shows the bright interiors of the refurbished RMLs following the fitting of fluorescent lighting and the application of white paintwork above seat level between 1992 and 1994. *Capital Transport*

Entering Trafalgar Square in June 1998, RML 2732 displays a somewhat disproportionately-sized General fleetname, as if to make up for the ineffectual red fleetname central in the gold-coloured relief band in the front panel once occupied by ventilator grilles. The brash 'L' side panel advertising succeeds in its application but fails to complement the vehicle's designed aesthetic. *Stephen Madden*

Left Resplendent in its revived Green Line condition and livery, RMC 1461, perhaps confusingly to an older generation, does a turn on route 15, providing extra comfort for its passengers by means of the deeper seat cushions it still retains from its Green Line days. Stagecoach's normal livery treatment for its RMLs can just be seen on the bus overtaking it. *Stephen Madden*

London General's RML 2605 on route 14 is about to cross the Thames by means of Putney Bridge with the same owner's Northern Counties Palatine II-bodied Volvo Olympian NV 165 on the 74 tucking in behind in July 1998. Will the rear bus become a classic also? *Geoff Rixon*